Copyright © 2021 Spunky Science

All rights reserved. No part of this book may be altered, reproduced, redistributed, or used in any manner other than its original intent without written permission or copyright owner except for the use of quotation in a book review.

LAB RULES

Wash your hands

Use goggles

Do not eat or drink

Be responsible

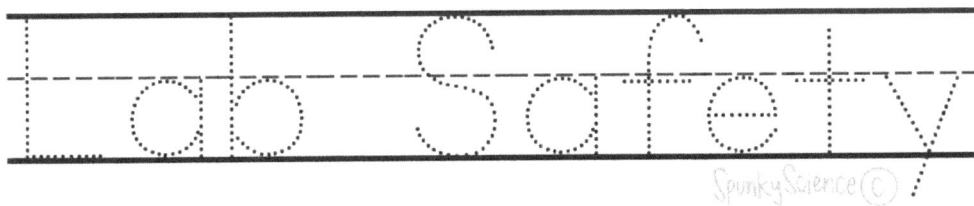

Minutes   Seconds

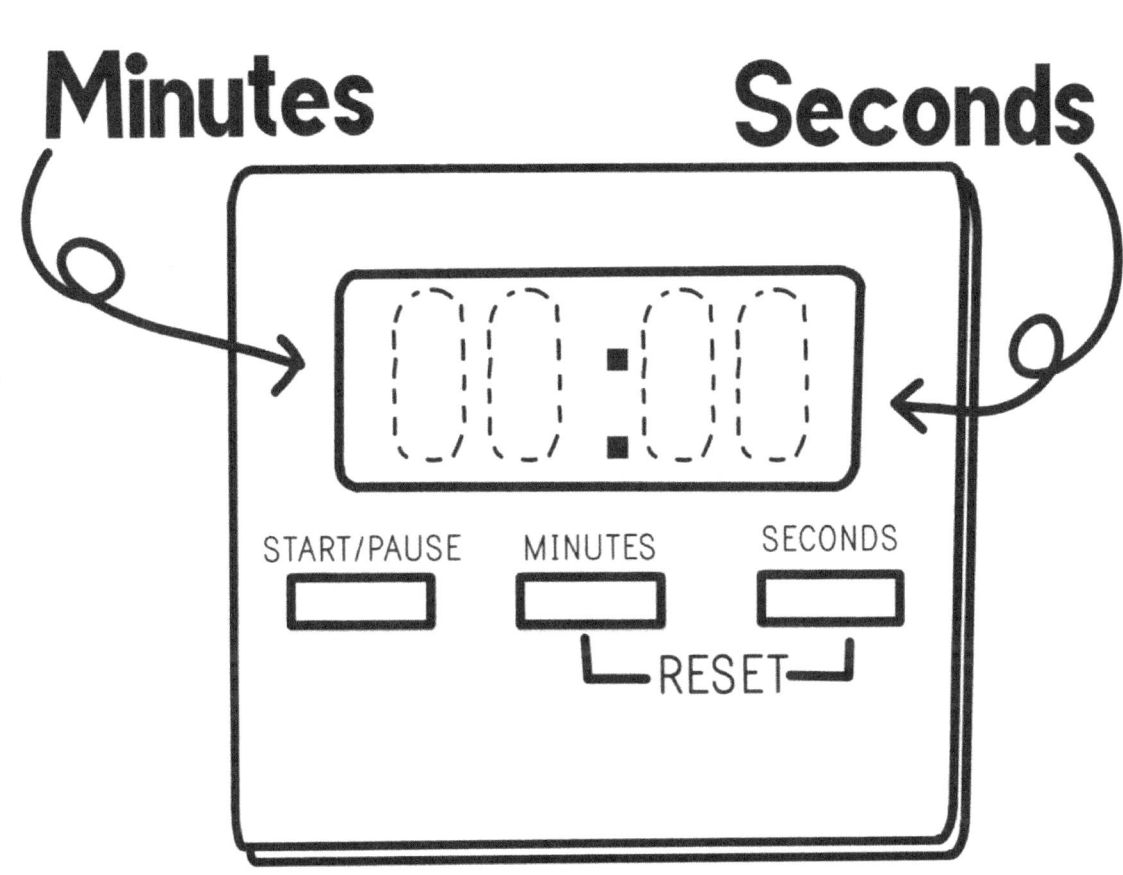

A timer is used to record how long something is happening. **This timer measures time in minutes and seconds and has a pause** button.

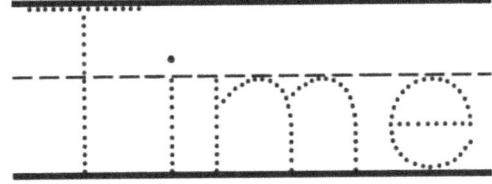

RAIN GAUGE

A rain gauge measures how much rain falls in a specific area.

Rain Gauge

RULER

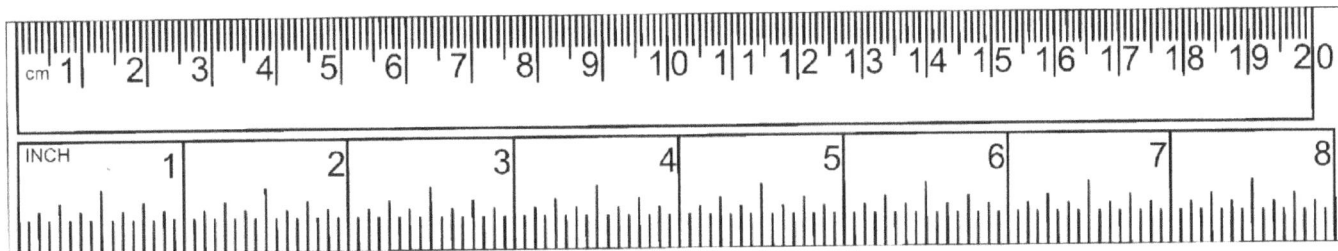

A ruler is a tool that is used to measure the length of an object. Each side measures in a different unit.

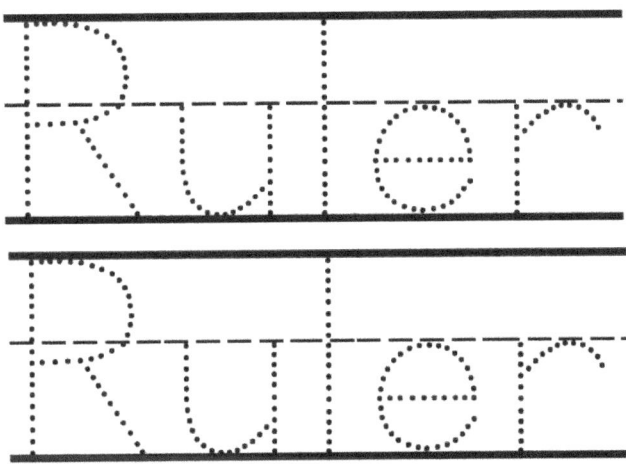

THERMOMETER

A thermometer measures how much heat something has. **If it has little** heat, **the number is lower and if it has a lot of heat** energy, **then the number is** higher.

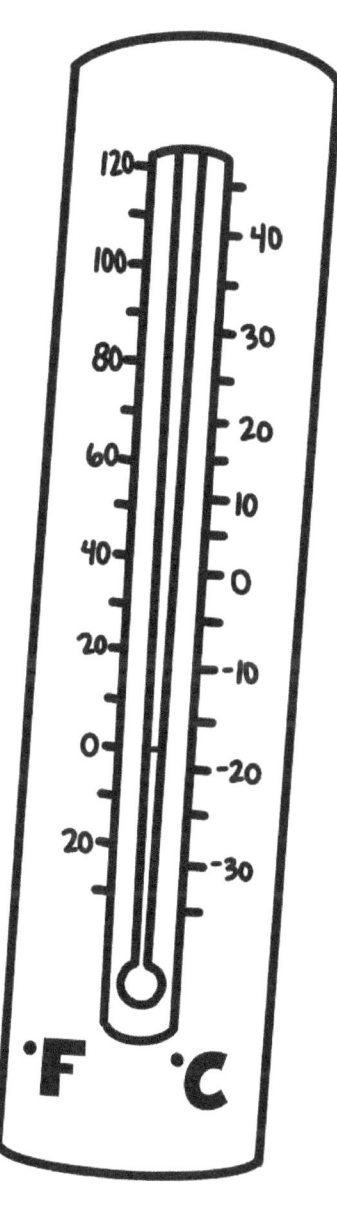

Some tell you the number while others you have to figure out.

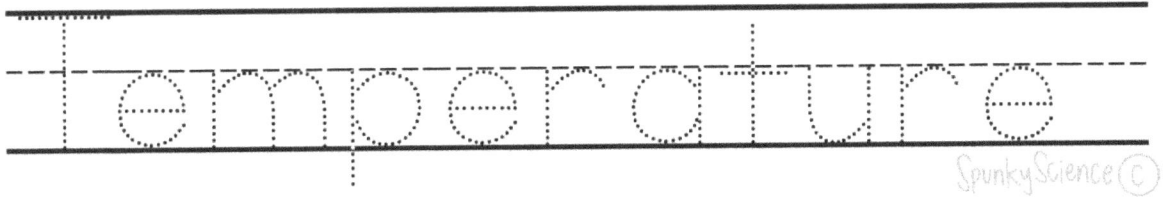

ENGINEERING DESIGN PROCESS

1. Identify the problem
2. Brainstorm solutions
3. Select a design
4. Build a model or prototype
5. Test and evaluate
6. Optimize the design
7. Share the solution

PUSH OR PULL

Pushes or pulls can have different strengths and directions

Rolling a ball

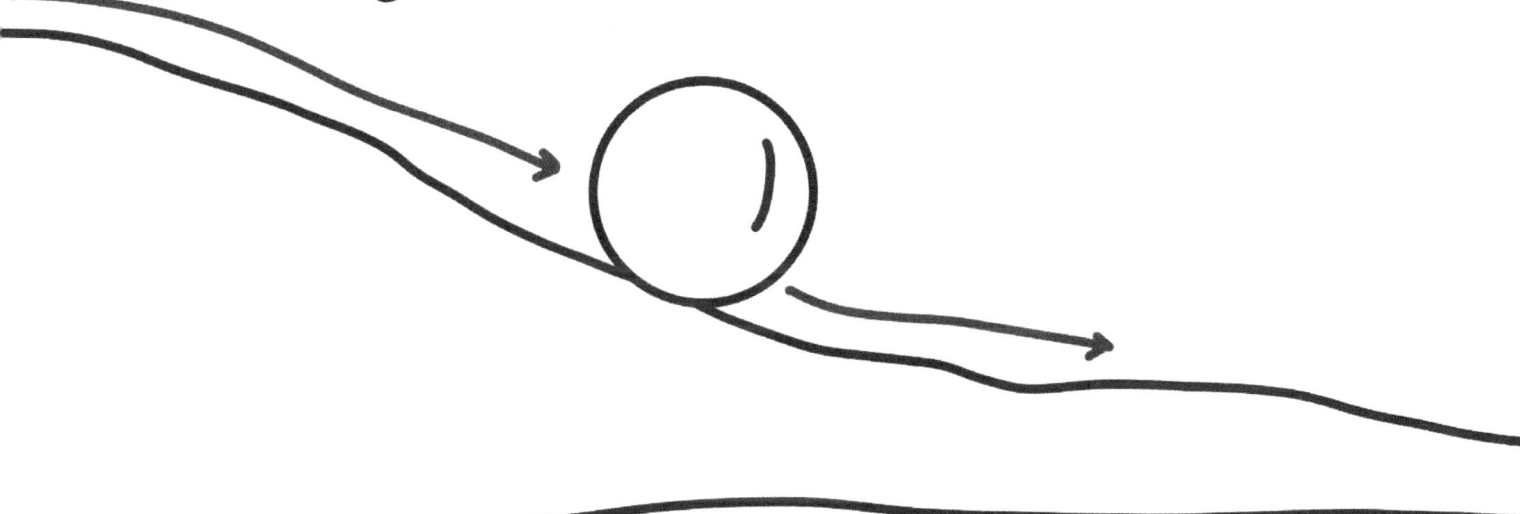

Pushing an object

Two objects colliding can change the force **and** direction **of the object**

When these two monster trucks collide, **both trucks will slow down and change** direction.

Sunlight warms Earth's surface

Material that is left in the Sun for a while is warmer compared to the same material kept in the shade.

Your school allows you to eat lunch outdoors, but it's too hot outside! Draw a diagram of a structure that would reduce the warming effect of sunlight on the area.

Draw your design:

What do plants need in order to survive?

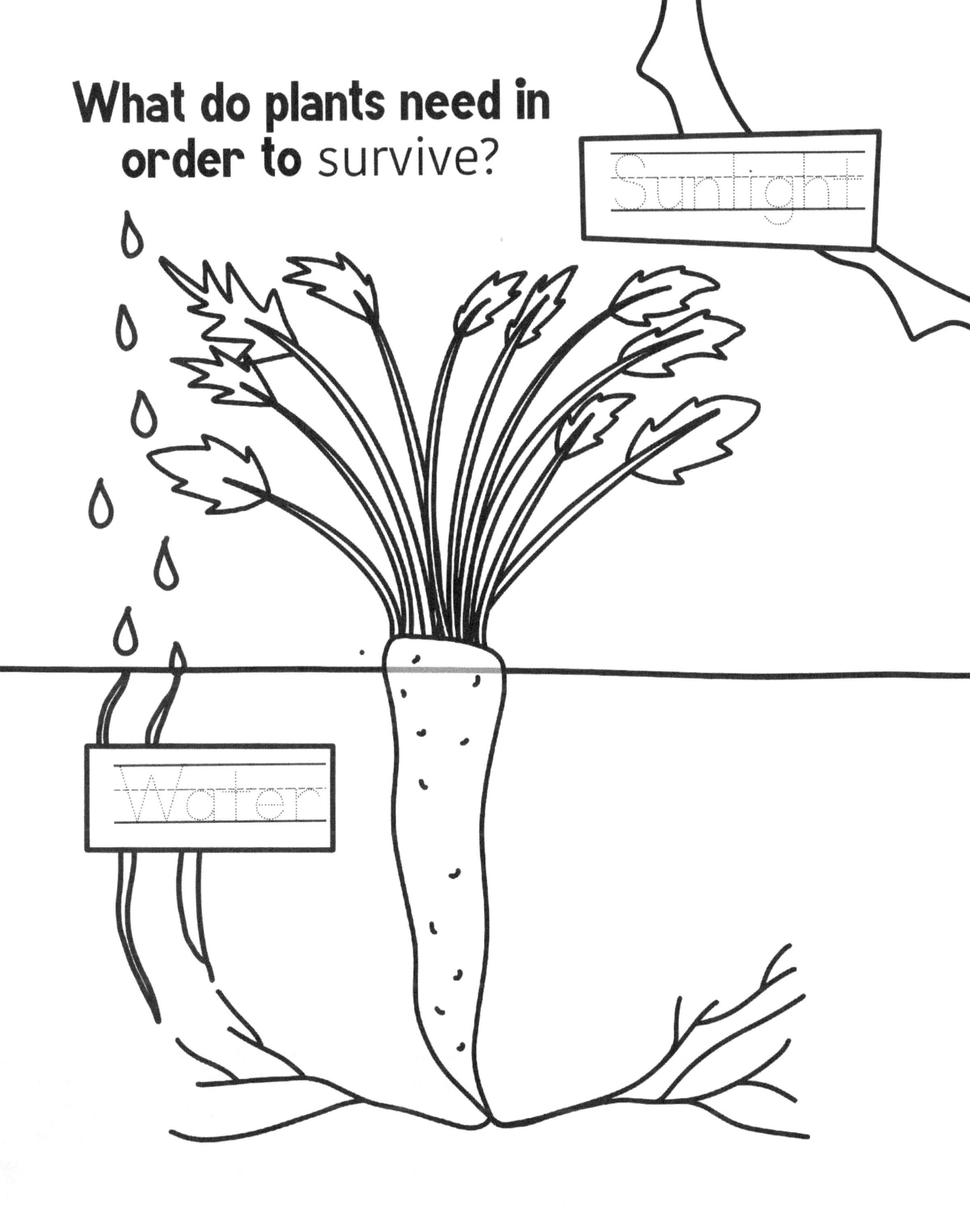

Sunlight

Water

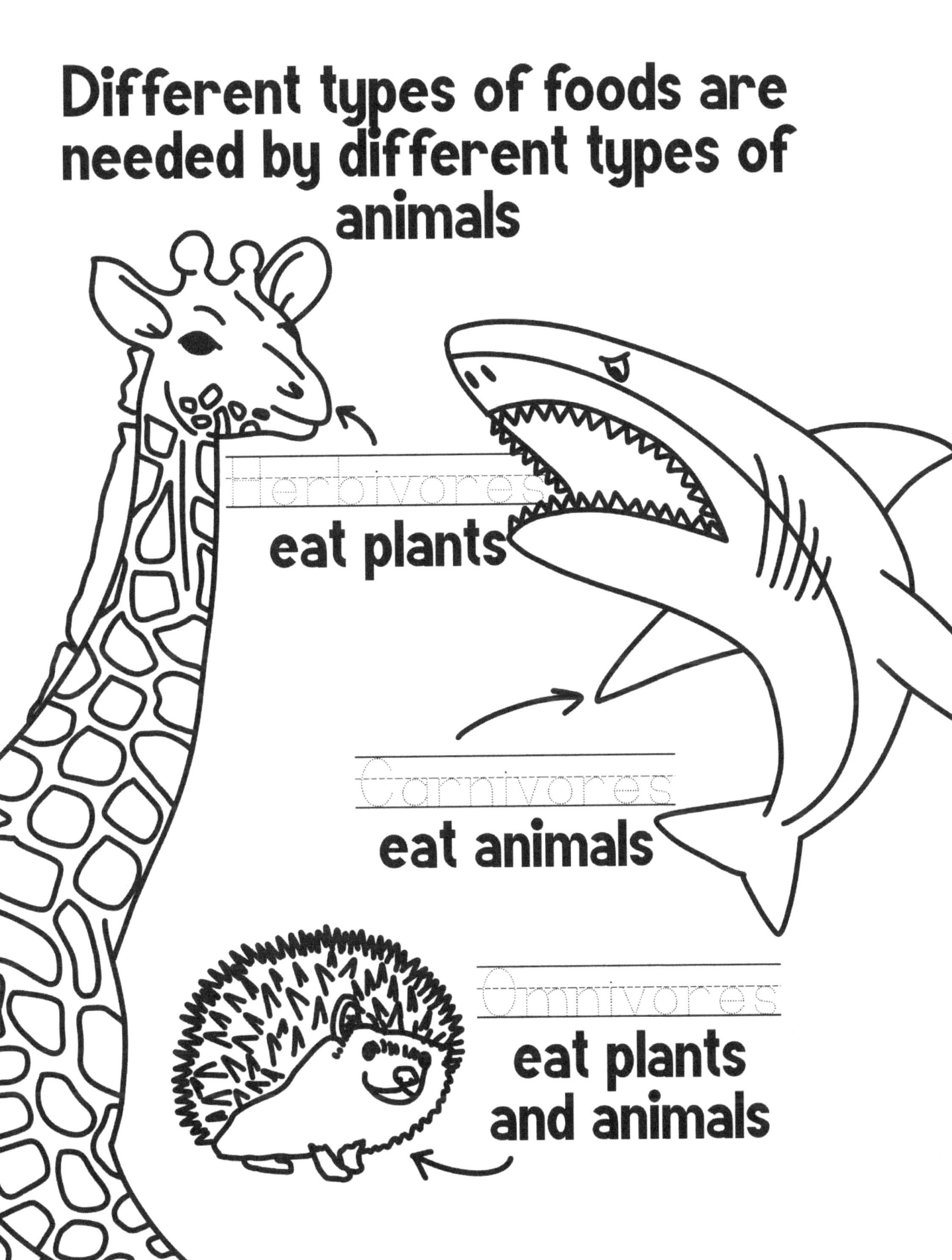

The needs of different plants and animals determine the places that they live.

Deer eat buds and leaves, so they live near the forest.

Grasses need a lot of sunlight so they are found in meadows

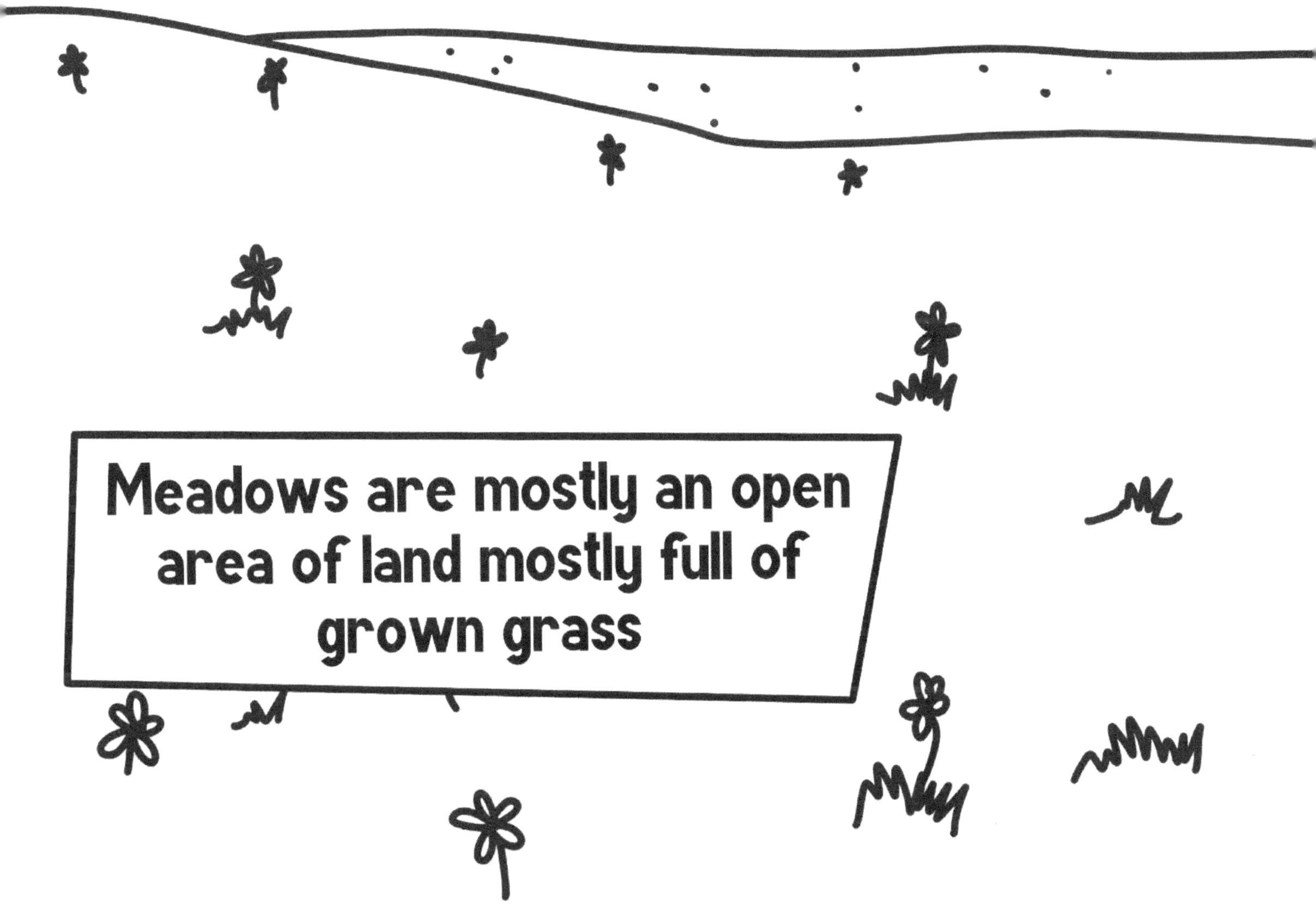

Meadows are mostly an open area of land mostly full of grown grass

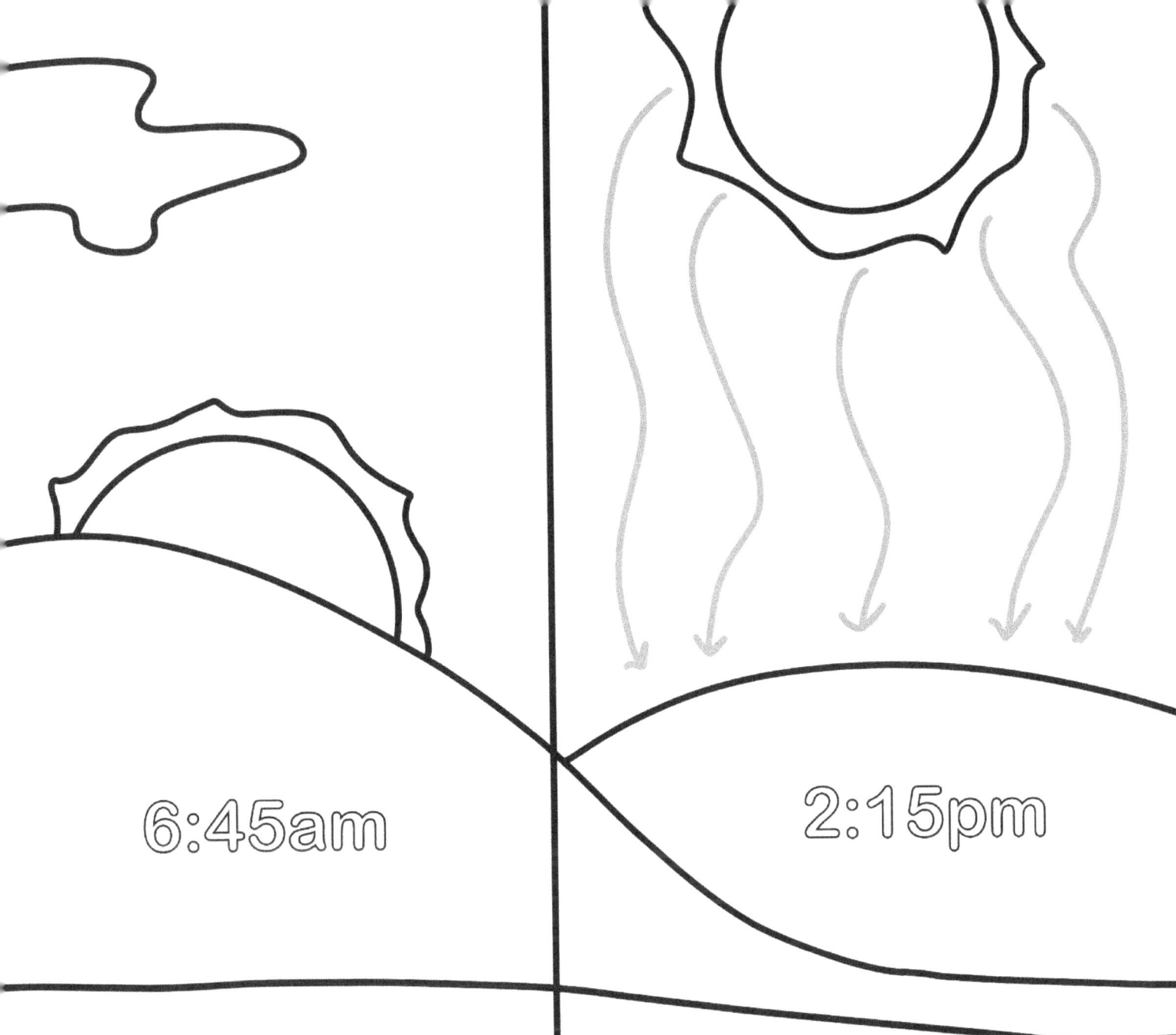

WEATHER FORECAST

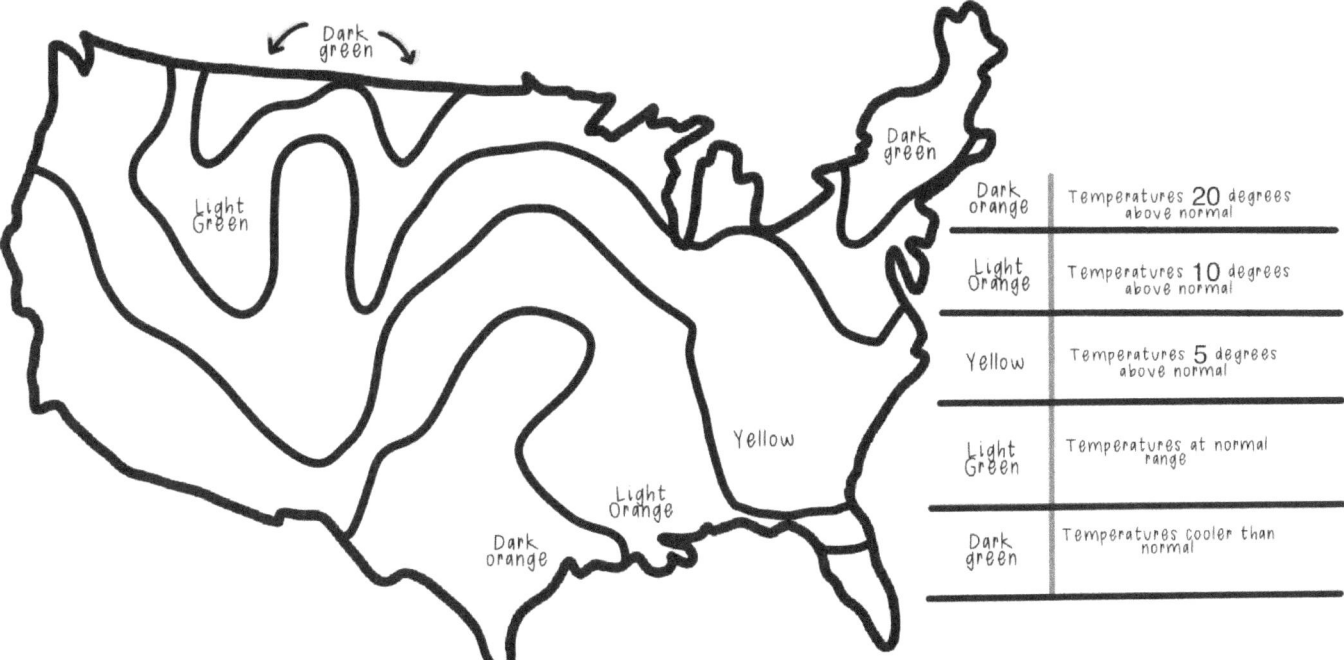

Dark orange	Temperatures 20 degrees above normal
Light Orange	Temperatures 10 degrees above normal
Yellow	Temperatures 5 degrees above normal
Light Green	Temperatures at normal range
Dark green	Temperatures cooler than normal

Weather forecasts help people prepare for and respond to severe weather.

Weekly Weather Update

Monday	Tuesday	Wednesday	Thursday	Friday
75	72	74	78	82
Sunny Warm	Cloudy	Cloudy Windy	Windy	Sunny Warm

WEATHER TOOLS

Weather is the way the air and the atmosphere feels. You can measure weather by a few different tools.

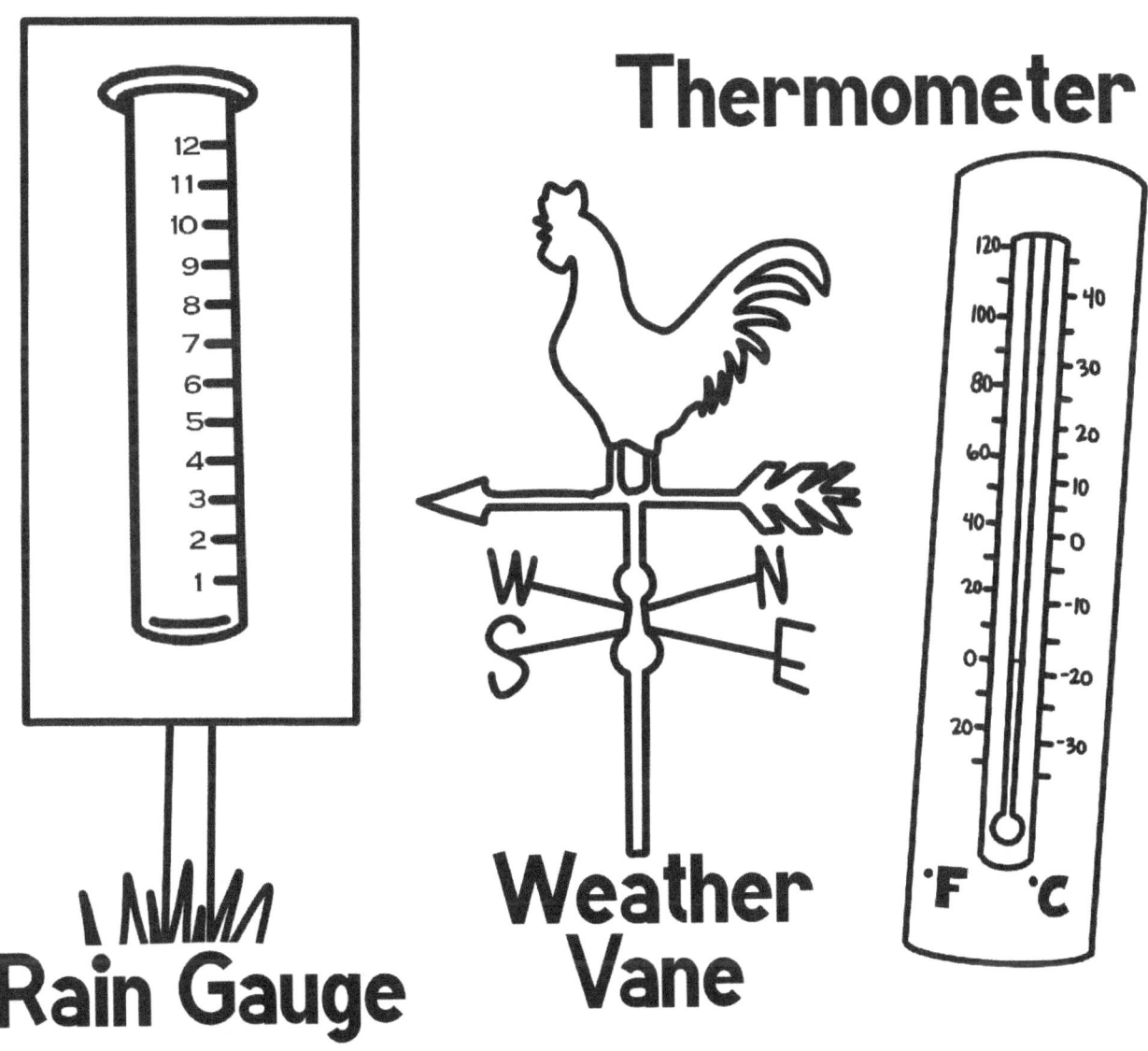

Rain Gauge

Thermometer

Weather Vane

PREPARING FOR SEVERE WEATHER

When you are preparing for severe weather, **you need to make sure that you have certain things packed and** ready!

Car full of gas

Flashlight and Batteries

Canned Food and Can Opener

Bottled Water

CLOUDY

Cloudy weather is when the sky is covered, **or mostly** covered, **in** clouds

Let's **figure out the number of** sunny, windy, rainy, **warm days in one** month!

Sunday	Monday	Tuesday	Wednesday	Thursday	Friday	Saturday

Draw these symbols on the days that shows that type of weather.

Sunny — Cloudy — Windy — Warm

Let's count! Write the number of days that were sunny, windy, rainy, and warm this month in the spaces below.

_____ **Sunny**   _____ **Windy**
_____ **Cloudy**   _____ **Warm**

HUMAN IMPACTS ON LAND

Impact: Cutting trees to produce paper

Solution: Reusing paper
Recycling Cans
Recycling Waste

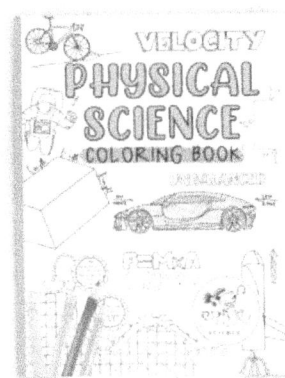  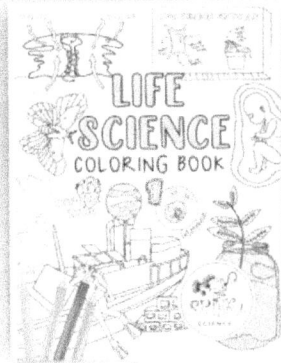 

www.ingramcontent.com/pod-product-compliance
Lightning Source LLC
Chambersburg PA
CBHW081103240526
45465CB00026B/3288